INTERNATIONAL CENTRE FOR MECHANICAL SCIENCES

COURSES AND LECTURES - No. 111

Sir JAMES LIGHTHILL
UNIVERSITY OF CAMBRIDGE

PHYSIOLOGICAL FLUID MECHANICS

FREE LECTURE
OCTOBER 1971

UDINE 1971

SPRINGER-VERLAG WIEN GMBH

This work is subject to copyright.

All rights are reserved,

whether the whole or part of the material is concerned

specifically those of translation, reprinting, re-use of illustrations,

broadcasting, reproduction by photocopying machine

or similar means, and storage in data banks.

© 1972 by Springer-Verlag Wien

Originally published by Springer-Verlag Wien New York in 1972

ISBN 978-3-211-81133-7   ISBN 978-3-7091-2963-0 (eBook)
DOI 10.1007/978-3-7091-2963-0

PREFACE

The first lecture is in the nature of a general survey of fluid flows within the human body: including the lungs (airflow in the airways and the special characteristics of the pulmonary blood circulation), the general systematic circulation of the blood, the kidneys and the urinary tract. Problems of the microcirculation, including blood flow in the narrower capillaries, gas exchange with the terminal airways (alveoli), and exchange of gas and nutrients with peripheral tissue, and postponed to the more specialised second lecture, which describes in some detail modern views concerning peripheral resistance.

Udine, October 1971

1.1. Introduction

Work in physiological fluid dynamics needs very close and intimate collaboration between specialists in physiological science and specialists in the dynamics of fluids. The necessary collaboration has to be preceded by a process of mutual education sufficiently prolonged to bring about on each side an adequate understanding of the other side's language and modes of expression, as well as recognition of which are the main areas where the other discipline has developed a particularly extensive and intricate body of knowledge and skills which can be called upon when required. After this, real communication between the different specialism becomes possible, and can lead to effective research progress.

In this lecture I shall indicate features of this collaborative research that have particulary struck me, as a specialist in fluid dynamics, during the six years in which I have been engaged in it with several colleagues among whom the leader on the physiological side was Dr Colin Caro. The first feature I want to emphasise is the richness of the field from the fluid dynamicist's point of wiew. This will not, I believe, come as a surprise to anyone with even an elementary knowledge of the complex organisation of the human body and in particular of its respiratory tract and cardiovascular system, to say nothing of other flow systems like those for transporting lymph or urine. It is easy to believe that wide experience in the very extensive and long established discipline of fluid dynamics can help someone to make a contribution to research in these fields, after he has acquired enough elementary anatomy and physiology to be able to understand the words that his colleagues are using when they put problems to him.

The second feature I want to emphasise is of an exactly opposite character. Many of the most important problems in this field, even if the fluid dynamics in them is considered in isolation, are found to raise questions which, in the whole preceding history of research on fluid dynamics, have totally failed to be answered, or in some cases even to be asked! It is quite humbling to notice how often those questions in fluid dynamics, suggested by study of a physiological problem, turn out to be questions never tackled during all the vast development of knowledge in fluid dynamics aimed mainly at engineering applications. From another point of view, however, this adds richness to the field: it is by no means merely advice that the fluid dynamics specialist needs to give; constantly he is forced to initiate quite new researches on basic problems of fluid flow, and many of these are as interesting as the most _interesting_ investigations in the field that have been suggested by engineering needs.

You may ask "why do fluid flows within the hu-

Introduction

man body raise problems so very different from those reised by the huge assemblage of engineering fluid flows?" To this question I can attempt an answer in the form of a list of five main reasons

(i) <u>Unusual range of Reynolds number</u>: both in the larger airways of the lung and in the larger arteries the Reynolds number takes values from a few hundred to several thousand, and in this range the study of internal flows has been largely neglected – particularly at the high-Reynolds-number end of the laminar-flow regime. Engineers have concentrated mainly upon fully developed turbulent flows (except in relation to high viscosity liquids, where they have been interested in quite low Reynolds numbers); in the body, by contrast, only very sporadic or highly localized "bursts" of turbulence occur (Section 10). On the other hand, laminar internal flows gain greatly in complexity at these high Reynolds numbers because an enormously long so-called "entry region" is required to establish any simple flow field like Poiseuille flow in a straight tube (Section 3), and also because centrifugal action in curved tubes produces intense secondary-flow effects (Section 2)

(ii) <u>Unusual multiplicity of tube branchings</u>: both the lungs and the cardiovascular system are extremely intricate and complex branched networks of tubes, whereby the initial flow, of air entering the trachea (windpipe) or of blood leaving the heart, is subdivided after perhaps 20 to 30 separate branchings into an enormous number (many hundreds of million) of small individual flows, involving inflation of an alveolus to a diameter of a few hundred microns, or passage of blood through a capillary of diameter less than ten microns. In the earlier stages of this branching process the flow pattern suffers a major distortion at each bifurcation, from which, for reasons given under heading (i), it may well not recover before the flow branches yet again (Section 7). In the later stages, where the Reynolds number is too small for this difficulty to arise, the nature of the flow becomes hard to analyse owing to the vast number of tubes in parallel and the insufficiency of information on distribution of numbers by length and by diameter.

(iii) <u>Unusual distensibility properties of containing vessels</u>: the networks of vessels containing both the airflows and the blood flows exhibit rather complicated distensibility relations. For example, the way in which a given degree of muscular action to expand the chest case increases the volume of different parts of the lung is subtly influenced by the ways in which they are separately deformed by the action of gravity on that very flexible structure. Again, the distensibility of the arteries is of great importance in matching the heart pump's reciprocating action to the steady perfusion of the peripheral capillaires, but it is complicated by viscoelastic features, which play a significant role in attenuating the pulse wave as it travels outward from the heart, and by nonlinear features due mainly to fibres of two very different materials in the arterial wall (elastin and collagen) being involved in resisting different degrees of arterial distension, not to speak of the effect of smooth muscle linings possessing various important control functions, particularly (in the case of the arterioles; see Lecture 2) control of the rate of perfusion of different peripheral tissues.

(iv) <u>Unusual fluid properties</u>: these are most marked in the case of blood, a suspension of some 40 to 50 per cent by volume of small deformable bodies, mainly the red blood cells which are highly flexible disk-shaped bodies of diameter about 8 microns, in a colourless fluid, the blood plasma. Although the plasma itself has flow properties close to the familiar ones described by Newton's viscosity law. The application of viscometry to whole blood yelds so-called values of effective viscosity that show a substantial increase with decreasing rate of strain, due to increased formation of various aggregations of red blood cells at the lower strain rates. Actually, data at the higher strain rates occurring in arteries indicate that errors from treating blood in those as a Newtonian fluid with constant viscosity may not be too serious. In the meantime some excellent research on the flow properties of suspensions is in progress; although this does not particularly encourage us to use different continuum models of blood with unusual rheology, it does emphasise how in small vessels with diameters a few times that of a red blood cell the migration of cells may produce a most unequal distribution of concentration across the tube, with (see Lecture 2) quite a complicated resulting effect on resistance. Incidentally, the air we breathe is also, regrettably enough, a suspension of dust particles and one of the key problems in respiratory fluid mechanics is to study what determines the proportion of particles of different sizes which the airflow may cause to be deposited at different levels in the bronchial tree.

(v) <u>Unusual pulsatility</u>: the regular flow reversals in pulmonary inspiration and expiration need careful study, e.g. to determine those departures from exact reversibility at different levels which alone can permit particle deposition. In the circulatory system of normal subjects, valves prevent the total blood flow integrated across a cross-section becoming negative, although some reversed flow localised near an arterial wall may occur. Furthermore, the variation of integrated flow against time in the ascending aorta, where the systemic blood flow leaves the heart, takes the form of a strong surge lasting for less than half of the cycle followed by a far weaker flow during the remainder. If this pulse form is Fourier-analysed the first few harmonics are comparable in amplitude to the steady state component. It is tempting to study the propagation of different Fourier components of the pulse separately and assume that they can be linearly superposed, but evidence for significant nonlinear interaction between such components is accumulating: this can arise from the 'entry region" phenomomena and from the "non linear elasticity" phenomena noted under headings (i) and (iii) respectively. On the other hand, attenuation of the pulse wave makes such pulsatility far less impostant in the microcirculation and in the veins.

In the remainder of Lecture 1, I indicate several lines of research in progress that arise from these various difficulties and that happen to interest me personally, as follows.

1.2. Steady Secondary Flows

Centrifugal forces acting as a result of curvature of the streamlines of a "primary flow" produce "secondary flow" velocities at right angles to those streamlines, with motion of central fluid rowards the outside bend and a return flow towards the inside bend near a wall. The effect, which occurs both in curved tubes and at bifurcations, is strongest in the high-Reynolds-number laminar-flow regime in which it has been least studied. We know now (Barua 1963, McConologue & Srivastava 1968) that flow in a curved pipe in this regime under the action of a steady pressure gradient involves practically uniform motion towards the outside bend of the central fluid, which is being accelerated by the pressure gradient, whereas the return flow, retarded by viscous stress, is confined to a relatively thin boundary layer. This seems to explain (Lighthill 1970, p. 114) why the critical Reynolds number for transition to turbulence increases from a typical 2000 for straight pipes to a typical 6000 for curved pipes; these values are based on pipe diameter, whereas the Reynolds number based on boundary-layer thickness changes far less. In the meantime it was shown experimentally by Caro (1966) that these secondary flows do act to promote lateral mixing of injected substances, mixing which for substances of low diffusivity in the absence of turbulence would normally be very slow.

1.3. Entry Regions

For high-Reynolds-number steady flow in a straight pipe, it is well known (for an excellent recent review see Fargie & Martin 1971) that the length of the "entry region" required for an approximately uniform initial velocity distribution to come within 5% of a Poiseuille distribution is about $(0.03R)d$ where R is Reynolds number based on the diameter d and initial velocity U_0. This expression suggests that any one of the larger arteries is almost all entry region, where it is certainly not correct to apply the law of Poiseuille (although he was a physiologist!). Note that the steady pressure drop in an entry region exceeds the Poiseuille value by an amount often referred to as "the kinetic energy correction"; it is required mainly to accelerate the central flow to the more peaked Poiseuille distribution with its mean kinetic energy higher by $\frac{1}{2}\varrho U_0^2$ but the total correction, allowing also for extra resistance in the extra thin boundary layer near entry, is about $0.67 \varrho U_0^2$. When we consider the total pressure drop across a system with very numerous bifurcations like the arterial system, we can find that even though a single kinetic-energy correction is small in relation to that total, the summation of a large number of them in series may not be negligible.

Work to combine these last two topics is in progress, mainly because the very first entry region in the systemic circulation is itself a highly curved tube, the great arch of the aorta. We cannot expect a fully developed secondary flow in this entry region, into which the blood probably enters with small initial vorticity if the aortic valve opens properly, so that the vorticity of both the primary and secondary flows may be confined to an entry-region boundary layer. The central flow would then have the characteristic irrotational property of velocities greatest on the _inside_ bend (the opposite of that for a fully developed secondary flow), although secondary motions in the boundary layer would scour the _outside_ bend and cause the rates of shear and of mass of transport to be greatest there. Preliminary work by associates of myself and Dr Caro (especially Dr Scarton on the theoretical side and Dr Seed on the experimental side) is tending to support this rather simple description.

1.4. Incipient Atheroma

You may wonder why so much emphasis is being placed upon precise distributions of flow velocity or of rates of shear and mass transport. There are several reasons, but one of the most important is in relation to the onset of atheroma, a degeneration of the arterial wall which leads to so called hardening of the arteries. Actually the rates of shear and of mass transport at a wall have generally similar distributions in flow systems, and there are two exceedingly interesting lines of research which link these distributions with the distribution within the circulation of sites of incipient atheroma. Unfortunately the two lines of research, described for example by Fry (1968) and by Caro, Fitz-Gerald & Schroter (1971), are emphasising opposing trends. The Ciba Foundation has arranged a symposium in 1972 where I have to act as an impartial charman and try to help reach a consensus regarding the relevance of the two types of work, so I must be careful not to prejudge the issue. I will say on the one hand that Fry (1968) is able to show that extremely high rates of shear are able to produce actual mechanical damage to the endothelial lining of the arterial wall, and that this can foster the onset of incipient atheroma. On the other hand the claim of Caro, Fitz-Gerald & Schroter (1971) is that these very high rates of shear, required for mechanical damage, are not found in the actual circulation. They study the distribution of sites of incipient atheroma in large numbers of human and canine subjects and appear to exhibit a close correspondence between these sites and sites where there are low rates both of shear and of mass transport that is, of transport of some substance away from the arterial wall due to scouring by the flow. They have a complicated explanation of atheroma onset in such "dead-water" and other low-shear regions in terms of a failure of the blood flow to transport away from the arterial wall a sufficiently high rate the cholesterol synthesised inside it. Without expressing any personal view on these two extremely interesting and extensive research programmes, I will remark that since atheroma is one of the big killing diseases of modern society they both in their different ways suggest how important it will be to abtain a good knowledge of the distribution of rate of shear in the circulatory system.

1.5. Distribution of Shear in Branched Systems

We may continue for just a little longer with comments based on steady-flow fluid dynamics; in fact, by studying steady flow through a branched system. The relation of such studies to real flow in the cardiovascular system or in the lung airways may be that they can indicate typical distributions at instants of <u>local maximum</u> flow, where effects of rate of change of flow are smallest. Alternatively, for the blood flow, they may indicate something of the distribution of the <u>steady-state component</u> of velocity (except in so far as this may non-linearly interact with higher Fourier components).

The rate of shear at a particular point in a particular tube of a branching system in steady flow (Caro, Fitz-Gerald & Schroter 1971) depends on two factors: (i) a flow factor depending on the total flow Q in that particular tube; this factor is $32Q/\pi d^3$, the wall shear for simple Poiseuille flow at volume flow rate Q in a tube of diameter d : (ii) a geometrical factor depending on position in the tube, and representing any local departure of the rate of shear from its Poiseuille value. These departures take forms already discussed: thus in any entry length (Section 3.) the geometrical factor is considerably greater than 1 upstream, but it decreases downstream; in a curved tube (Section 2.) it is greater on the outside bend than on the inside; at a bifurcation, the geometric factor is large on the central "flow divider" where a new boundary layer tends to form, but is small on opposite parts of the tube walls, which are effectively "inside bends" for the flow curving round into the daughter tubes.

To these remarks I will add brief comments on

how the flow factor varies in a branching system, using the concept of the area ratio β at a bifurcation; here, β is the ratio of the combined area of the daughter tubes to the area of the mother tube. This area ratio is important both in the cardiovascular system and in the lungs, because it is necessary to achieve a large <u>total</u> increase in area through both these networks: two orders of magnitude from the aorta to the whole assemblage of systemic capillaries, and still more from the trachea to the alveoli.

If a fluid dynamicist were asked to design such a branching system, he would accept the need to make β consistently greater than 1 at each bifurcation, but he would be very cautious about making β considerably greater than 1 in at any rate the earlier bifurcations. This is because the mean flow velocity will be <u>divided</u> by β, and this retardation may cause flow separation if β is greater than about 1.2. Such flow separation may be undesirable either (i) because it increases resistance or (ii) for some other reason such as a postulated harmful effect of a "dead-water" region upon the arterial wall.

However this may be, the measured values of β at early stages of both networks are small: less than 1 for the main aortic bifurcation (Caro, Fitz-Gerald & Schroter 1971) and not more than 1.2 for other bifurcations. We note the consequences of this here in only the simple special case of a symmetrical bifurcation, with half of the mother-tube flow going into each daughter tube.

Then the main velocity is multiplied by β^{-1}, the tube diameter by $\left(\frac{1}{2}\beta\right)^{1/2}$, and the Reynolds number by $(2\beta)^{-1/2}$. This reduction in Reynolds number is important because after enough bifurcations the Reynolds number will be reduced to values of 10 or less for which flow separation is not expected even for quite large values of β, which there would become acceptable. Finally, the "flow factor" in the rate of shear, namely $32Q/\pi d^3$, is multiplied by $(2/\beta^3)^{1/2}$. We now see that values of β in the early bifurcations are small enough so that this quantity is greater than 1, and conclude therefore that this "flow factor" increases downstream. In general, then, rates of shear are least in "proximal" arteries (those nearest the heart) and, within them, in those sites where the "geometrical factor" is smallest.

1.6. Distribution of Resistance in Branched Systems

We note also how the different parts of a branched system may contribute to the total pressure drop in a steady flow through it. To do this we make the crude assumption that as tube diameters are reduced the associated lengths of tube are reduced approximately in proportion, which appears to be broadly speaking true.

Distribution of Resistance in Branched Systems

The pressure drop predicted by Poiseuille's law exhibits a simple behaviour: to obtain the value in the daughter tubes, we multiply the value in the mother tube by the same factor $(2/\beta^3)^{1/2}$ as for the shear. We need to remember that the true pressure drop will exceed that predicted by Poiseuille's law in entry-length regions, and for other reasons at high Reynolds number (see Section 7 below, and also heading (i) above), but can immediatly observe that while the shear is increasing distally (that is, towards the periphery, and for the reason that β does not much exceed 1) the Poiseuille contribution to pressure drop will also increase.

The cardiovascular system and lung airways differ in the details of this. It appears that the former the values of β are thus kept down for very many generations; certainly to tube diameters less that 0.1 mm. This is consistent with the 'fluid dynamicist's plan' to wait until Reynolds number is small before allowing higher values of β that might otherwise permit flow separation. By contrast, in the lung the ratio $(2/\beta^3)^{1/2}$ falls below 1 soon after the fifth generation, where Reynolds number is still several hundreds. Substantial flow separations in the immediately subsequent generations can be expected but for various reasons may be less harmful in the airways than in the blood-stream.

Associated with this is the fact that in the blood-stream the greatest pressure drop occurs in small vessels of diameter less than 0.1 mm (some features of this are discussed in Lecture 2); whereas all the significant pressure drop in the airways network occurs in the first ten generations, with diameter exceeding 1 mm (Macklem & Mead 1966). Put in a different way, the arterial resistance is a low-Reynolds-number problem but the bronchial resistance is a high-Reynolds-number problem.

1.7. Bronchial Resistance

Accordingly, the relatively high-Reynolds-number considerations of Lecture 1 are relevant to pressure drop, and its enhancement beyond the Poiseuille prediction, mainly in the context of bronchial resistance. With this in mind, Pedley, Schroter & Sudlow (1970) studied that enhancement in detail for branching geometries typical of the first ten generations of lung airways, and came to some interesting conclusions.

They showed that the secondary flows present in a bifurcation are so powerful that the initial velocity distribution in a daughter tube is distorted in shape to a quite extraordinary degree, with a very high peak near the inside bend; and that very little of this distortion is lost through the whole length of the daughter tube. In consequence, the ratio of viscous dissipation to that predicted by Poiseuille's law is _much_ more than in an ordinary entry length with uniform entrance conditions; in fact, it is $0.33\,(Rd/1)^{1/2}$ where Reynolds number R is based on diameter d and mean velocity, while 1 is tube length.

This means that the contribution to pressure drop of the first few generations where R is greatest is still further enhanced. Furthermore, it implies a total pressure drop in this high-Reynolds-number laminar-flow region proportional to flow to the $(3/2)$th power. Actually the flow in the trachea itself is normally turbulent, but this is in a Reynolds number range around 10,000 where the turbulent resistance law happens to be also close to a $(3/2)$th power law! Pedley, Schroter & Sudlow (1970) conclude, therefore, that pressure drop through the _whole_ bronchial tree follows such a law, and describe some confirmatory experimental data.

1.8. Velocity Distributions in Pulsatile Flow

Some consequences of the pulsatile character of arterial blood flow, to which I must now urgently turn, will at first be considered without taking arterial elesticity into account. Probably this is not too bad an approximation if we are interested in how the pulsatile local distribution of velocity across a systemic artery responds to the pulsatile variations of local pressure gradient. (Note that typical variations of arterial diameter as a pulse passes are only about $\pm 2\%$; note also that the pulse wave velocity is at least five times the maximum blood velocity, so that if we neglect elasticity and thus make the effect of pressure changes propagate infinitely fast there is not such an enormous exaggeration involved!)

The velocity distribution associated with a single Fourier component of the blood pulse, with radiant frequency ω, depends critically on the quantity called α by Womersley (1955), which is a ratio of the tube radius $a = \frac{1}{2}d$ to the oscillating-boundary-layer thickness $(\nu/\omega)^{1/2}$ where ν is kinematic viscosity (say, $4\,mm^2 s^{-1}$ for blood flow in large arteries). With $\omega = 8s^{-1}$ (typical value for the first Fourier component in normal humans) this thickness is about 0.7 mm, while for the nth Fourier component this figure should be multiplied by $n^{-1/\alpha}$.

In the large arteries, then, α is of order 10 and the response to sinusoidal pressure gradient is predicted to be a velocity distribution uniform across the tube except for a boundary layer occupying about 10% of the tube radius. Almost all the pressure gradient goes into acceleration, that is, into combinating the inertia of the fluid, rather than into combating viscous resistance, so that the flow lags by almost 90° behind the pressure gradient (*). Within the boundary layer,

---

(+) This is true even though viscous resistance is as much as./.

however, the lag is less; only 45° for the rate of shear at the wall itself.

1.9. Pulse Propagation

The simple classical theory of pulse propagation in an elastic thin-walled tube with Young's modulus E and wall thickness h gives the basic value $(Eh/2\rho a)^{1/2}$ for the wave velocity c. Much study has been devoted to estimation of departures from this value due to fluid viscosity or due to complicated wall properties (thickness; compressibility; "tethering"; viscoelasticity; nonlinearity) but surprisngly little change in the result is predicted for the real arteries. Attenuation of the wave seems to be underestimated by effects of fluid viscosity alone, but significantly augmented by arterial viscoelasticity. Wave velocity may perhaps be very slightly greater for the first Fourier component (with reduced inertia coefficient owing to the thicker boundary layer) than for higher ones.

Measurements show that typical wave velocities

./. twice the Poiseuille prediction (owing to the thin boundary layer) for $\alpha = 10$. It is only for $\alpha < 3$ that inertial resistance, which in this region (arteries of diameter $< 4mm$) is predicted to follow Poiseuille's law rather accurately.

increase peripherally to values around $10\,\text{ms}^{-1}$ from values for the aorta around $5\,\text{ms}^{-1}$, consistently with a much lower modulus of elasticity observed for the material of the aorta. These measurements are made by timing the passage of an artificially induced sharp pressure disturbance.

By contrast, the natural human pulse is <u>not pure</u>ly a wave travelling outwards from the heart. It includes a number of reflexions from bifurcations. The theory of wave <u>re</u>flexion at a bifurcation makes it depend on the "admittance" Y: this is the reciprocal of "impedance", and is the ratio of the amplitudes of the volume flow fluctuations and the <u>pres</u>sure fluctuations, being given for a travelling wave in a tube of cross-sectional area A as $Y = A/\varrho c$. A positive pressure pulse produces a <u>positive</u> reflected pulse (of reduced amplitude) at any bifurcation where the sum of the admittances of the daughter tubes is less than that of the mother tube. This is the case at the main aortic bifurcation (where we have seen that the total area actually decreases, while c increases) and at some other early bifurcations in the arterial tree.

This is why the waveform in the aorta and a few other large arteries includes a <u>positive</u> reflected-wave <u>compo</u>nent, and is thus intermediate in character between a travelling wave and a standing wave. These considerations make the largest arteries, and aorta in particular, able to act effectively as a so called <u>Windkessel</u>, storing by its capacitance the stroke volume from the cardiac output for delivery at a very steady rate through the peripheral circulation (see for example Taylor 1966).

Study of the transmitted wave at such a bifurcation explains also why pressure amplitude in one of the arteries in the leg may exceed, owing to reduced admittance, the pressure amplitude in the aorta. The smoothed form of the pressure-time curve at such a site may be attributed to greater attenuation for higher harmonics. The cause of another feature of it (lagging phase of the second harmonic, leading to a more marked "diastolic peak") is more in dispute; we note this as a point where either the influence of nonlinear effects or the greater wave velocity of the first Fourier component could produce an effect in the direction observed.

1.10. Turbulence in the Blood Stream

Uncertainly that has long existed about whether turbulence appears in the human aorta has now been resolved by work using rapid-response velocity-measuring probes in at least three centres (work of Dr Ling in Washington, Dr Schultz in Oxford, Dr Seed in London, together with several collaborators). The peak Reynolds numbers are such that the secondary flow in the arch of the aorta might under steady conditions just avoid the onset of turbulence (Section 2).

Actually turbulence appears for quite a small fraction of each cycle, while the velocity is first decelerating (see for example Seed & Wood 1971).

I am personally inclined to attribute this turbulence to change in the boundary-layer velocity profile, whose response very near the wall to the retarding pressure gradient has a phase advance of 45° (see Section 8) over that of the central core fluid. This produces the type of velocity profile with a point of inflexion, which is known to be enormously more unstable than any other type. "Bursts" of turbulence are commonly generated soon after such a point of inflexion appears locally near a wall. They may die away quite quickly, however, when the conditions change (see for example Lighthill 1970).

Turbulence is not common elsewhere in the circulation, except where jets form; for example, owing to valvular stenosis (failure to open completely); a circumstance that of course is of great clinical value. In the usual technique for blood pressure measurement using an arm cuff inflated to a known pressure, that is allowed to fall while the physician listens to the flow immediately downstream, the well known Korotkov sound is heard once the cuff pressure is intermediate between systolic and diastolic. This, however, is a more musical sound than that produced by turbulence.

Recently the Korotkov sound was rather clearly explained in excellent work of a combined theoretical and ex perimental nature as due to a rather regular nonlinear limit-cycle oscillation which occurs while a tube is in the process of collapsing under excess external pressure. Very roughly speaking, the mechanism for this is that the rate of flow constriction increases to a point where the fluid dynamics demands unsteady pressures which temporarily halt or even reverse that rate, which then begins to increase again because the excess external pressure once more dominates.

1.11. Urinary Tract

Flows of other fluids besides air and blood pose very interesting problems. In the kidney, remarkable mechanisms, including some similar to the chemical engineer's "countercurrent exchangers", form the urine out of a fraction of the blood plasma. This passes to the bladder through a ureter (one for each kidney) by means of a peristaltic pumping mechanism, whose mode of operation repays study for many reasons. An excellent review of peristaltic pumps has recently been given by Jaffrin & Shapiro (1971).

2.1. Introduction

This more specialised Lecture 2 is concerned with blood flow in vessels of diameter less than 0.1 mm (that is, 100 microns), and especially with flow in the arterioles, capillaries and venules, with diameters ranging from at most 50 microns down to values as low as 5 microns. This whole peripheral part of the circulation, called the microcirculation, was seen already in Lecture 1, section 6 to be where most of the drop in mean pressure between the arteries and veins, amounting to between 100 and 200 mb in normal human beings, occurs.

Lecture 2 is particularly concerned with quantitative aspects of peripheral resistance: that is, of the dependence of this pressure drop upon flow rate, a dependence which turns out to be interestingly nonlinear. Peripheral resistance is important because it determines the rate of working demanded of the heart, and also because the body's very needful ability to bring about large selective variations in the rate of perfusion of different parts of the periphery (with flow rates changing by an order of magnitude) depends on the power of the smooth muscle lining the arterioles to make large changes in peripheral resistance by expansion or contraction.

Another essential characteristic of the microcirculation is the function of the capillaries (vessels without a smooth muscle lining, with diameters around 10 microns or less) in various exchange processes. Capillaries in the lung are separated from alveoli by exceedingly thin membranes which readily admit exchange of O_2 and CO_2 between the air in an inflated alveolus and the blood in a perfused capillary. Within the blood, a similar exchange process occurs between the plasma and the haemaglobin, which of course is packed within the red blood cells surrounded by an even thinner cellular membrane. In the systemic capillaries the same exchange processes occur (though in the opposite direction) and in addition there is exchange of various nutrients between the plasma and the surrounding tissue. Some of this is by diffusion, and some by convection as small quantities of plasma (less than 1% of the total flow) squeeze through gaps (less than 0.1 micron wide) in the endothelial wall of the capillary (Guyton 1966). Much of this fraction of the plasma is ultimately sucked back into the low-pressure end of some capillary, but the rest is drained away through the quite separate lymphatic circulation, to rejoin the blood flow only in one of the largest veins.

      The feature which distinguishes the fluid flows in the microcirculation most markedly from the flows in larger vessels discussed in Lecture 1, and also from most flows that are generally familiar, is that fluid inertia is totally negligible. We are concerned with flows whose pattern is determined solely by the need for pressures gradients and viscous forces to be exactly in balance, because inertial effects (accelerations times densities) are too small to disturb that balance at all. Essentially this is because the Reynolds number is always considerably less than 1 in vessels of diameter less than 0.1 mm and falls as low as 0.001 in the narrowest capillaries, while also the Womersley parameter α (Lecture 1, Section 8), which measures inertial effects due to <u>pulsatility</u> against viscous effects, is considerably less than 1. In this world of the microcirculation, this "extreme-low-Reynolds-number world", where inertia is totally negligible, we have to forget most of what we learnt in studying the larger vessels and concentrate on some completely different considerations.

Let me list just what we ought to take pains to forget, in these corcumstances when pressure gradients are purely balanced by viscous forces. We must forget about Bernoulli's equation: there is no measurable difference between static and dynamic pressures. We must forget about centrifugal forces: fluid can now negotiate sharp bends without any difficulty at all, and without setting up any kind of secondary flow. Generally speaking, in fact, motions are much less sensitive to vessel geometry: there is practically no tendency to flow separation. Indeed, in a certain subset of cases (fluid satisfying Newton's viscosity law flowing in rigid vessels) the flow is completely "reversible".

The insensivity to geometrical detail is further shown by the virtual absence of any "entry region". If fluid satisfying Newton's viscosity law enters a tube at a large angle to the axis, for example, then the difference between the motion and a Poiseuille motion with the same total flow is reduced by an order of magnitude already in an axial distance of half a radius (Fitz-Gerald 1971): an exceedingly quick adjustment to Poiseuille flow. Study of the microcirculation is facilitated by these simplifications, and by the possibility of taking over ideas from parts of low-Reynolds-number fluid dynamics which were developed in an engineering context: for example, from the hydrodynamic theory of lubrication.

To be set against these advantages, however, is the major difficulty mentioned already in the introduction to Lecture 1, under heading (iv): we are constantly forced to take into account <u>interactions between the red blood cells and the wall</u> in any of these tubes whose diameters are at most a few times those of the red blood cells. Accordingly, much of Lecture 2 (from Section 4 onwards) is devoted to those special features of the microcirculation resulting from the fact that blood is a highly concentrated suspension of red cells (not to mention many other costituents, which however are probably far less influential upon its dynamics) in plasma.

2.2. Vasomotor Control of Peripheral Perfusion

A remarkable feature of the systemic circulation, commented on already in Section 1.1, is the vasomotor control of peripheral perfusion: that is, the power of smooth muscle to alter the diameter of small vessels in such a way as to vary flow rate through the local microcirculation by al least an order of magnitude. Although muscle is absent in the walls of the capillaries themselves, there is a large body of evidence that vasomotor control is exercised almost entirely in the smallest pre-capillary vessels with diameters substantially less than 0.1 mm.

It would be natural to suppose that, in a intricate branched system, control of local resistance must be exercised in that group of vessels which makes the biggest contribution to total resistance, and there is substantial evidence that this is the case. In Lecture 1, we noted regretfully under heading (ii) that the evidence on distribution of tube numbers by length and diameter is scanty, but indicated in Section 1.6 the way in which any such knowledge that exists can be used to estimate the distribution of pressure drop through a branched system on the assumption of Poiseuille's law. Kraemer (1967) carried out such a calculation which suggests that under conditions of vasodilatation about half the total pressure drop in the circulation occurs in precapillary vessels of diameter less than 50 microns (what in his classification are described as the arterioles and the smallest group of arteries); about a quarter occurs in larger vessels and about a quarter in capillaries. (Note that in this Section all "pressures" mentioned are corrected for the purely hydrostatic component so as to become "equivalent pressures at the height of the heart").

Such a conclusion, that even before vasoconstriction about half of the total resistance in the cardiovascular system is confined to vessels in quite a small size bracket (say, 15 to 50 microns), which indeed are the vessels within which vasomotor control of peripheral perfusion is thought to be exercised, is quite striking. Neverthless, Kraemer has to assume an extremely large vasomotor increase in the resistance of those vessels to explain the large observed changes in perfusion rates, since half of the unperturbed total resistance is not under control.

It is natural to try to probe this slight difficulty by considering the effect of departures from Poiseuille's law. One type of departure, the high-Reynolds-number type, discussed in Lecture 1, would accentuate the difficulty, but to an insignificant extent since it augments the contributions of only those rather wide tubes where quite small pressure drops are in fact found. Another type characteristic of the microcirculation, the Fahraeus-Lindquist effect discussed be low in Section 1.4, was actually allowed for in Kraemer's calculations by assuming a 40% reduction in pressure drop below the Poiseuille value in tubes of diameter less than 25 microns. More recently another possibility was suggested (Lighthill 1968, p. 116), that a nonlinear resistance property of the microcirculation, resulting from features of the flow in the narrowest capillaries, may augment the effectiveness of vasomotor control; these features will be described in detail in Section 2.5.

It would be out of place to discuss here the many different types of stimuli that produce vasomotor action; most of them (for example, temperature as stimulating vasodilatation, or a wide variety of chemical stimuli) have little to do with fluid dynamics. It is however appropriate to note that marked vasomotor response to a rapid change of internal fluid pressure was established by Caro, Foley & Sudlow (1970). Their work suggests that a sudden pressure reduction, due to deflation of a cuff that had been producing congestion in the veins of the forearm leading to an excess venous pressure of 40 mb, may produce, after a delay time of 2 to 3 seconds, a sudden vasomotor increase in forearm blood flow to almost three times the normal resting value although during the brief delay before vasomotor action occurred the flow rate was above that resting value (possibly due to passive dilatation by the excess pressure) by only 25%. This example of an active response to pressure change emphasizes the need for the specialist in the mechanics of fluids and of passively deformable tubes to approach the study of physiological phenomena only with the greatest respect and caution!

2.3. Pulmonary Perfusion and Ventilation

The microcirculation in the lungs exhibits several features that contrast very sharply with those for the systemic micro circulation summarised in Section 2.2. There is no vasomotor activity in pre-capillary vessels, or elsewhere, and most of the resistance to flow appears to lie in the capillaries themselves (Fowler, West & Pain 1966). However, perfusion of the capillaries can be extremely unequal as between different parts of the lung, though we shall see that this is essentially a passive effect in which gravity plays a leading role. Since the lung's effectiveness as a gas-exchange system depends upon a reasonably good match being reached between the distributions of perfusion and of ventilation across different parts of it (Farhi 1966), it is worth observing that gravity also acts to promote inequalities of ventilation of the alveoli which go some (though not all) of the way towards matching the inequalities of perfusion (West 1966).

Typical capillaries in the lung pass through the thin septum that separates two alveoli, and are themselves separated from alveolar gas by the still thinner pulmonary membrane.

Accordingly they are very easily distorted whenever the alveolar pressure p_A exceeds the blood pressure in an adjacent capillary. This distortion is found by West, Glazier, Hughes & Maloney (1969) to involve very large reductions in mean width, from 6.5 to 2.3 microns; even for a fluid satisfying Newton's viscosity law this would involve an enormous increase of resistance, and consequent flow reduction, but for real blood, for reasons (Section 2.5) involving the size of individual red cells, it brings about a complete stoppage of flow.

It is necessary to clarify immediately the meaning of the two pressures here compared, starting with the alveolar pressure p_A. In lung inflation, involving expansion of hundreds of millions of alveoli to a diameter of a few hundred microns, an enormous increase of essentially wet surface area occurs; it is like the formation of a foam! Thus, a very large effort would be required for inflation if the liquid wetting the surface were water with its high surface tension. Fortunately a certain protein is present in the liquid which operates rather like the detergents used to facilitate foam formation; this enormously reduces surface tension (Pattle 1966). It means also that the gas pressure p_A in an alveolus is essentially what determines closure of an adjacent capillary, any surface-tension correction to it being rather insignificant.

As for the blood pressure in the adjacent capillary, this must take values intermediate between the pressures

p_a and p_v on the arterial and venous side of the pulmonary circulation, the pressure fall between these values being balanced by capillary resistance. The heart pump produces a mean excess pressure in the pulmonary arteries less by an order of magnitude than that in the aorta: for example, about 12 mb during resting, although during exercise this can easily be doubled. Such figures are comparable with the hydrostatic pressure difference between the top and bottom of the lung, and in this Section we cannot simplify as we did in Section 2.2 by ignoring the hydrostatic component, since pressures in the circulation are to be compared not merely with each other but also with the gas pressure p_A in an alveolus.

Thus p_a and p_v show a hydrostatic decrease with height of 1 mb per cm . Actually there is some decrease in the alveolar pressure p_A with height , at a rate around 0.3 mb per cm. Fully to relate this to the mechanics of the lung hanging under gravity, supported through the thin layer of pleural fluid, requires lengthy discussions outside the scope of fluid dynamics; but the outcome is that the highly deformable lung structure essentially has some properties in common with a fluid, including just such a gradient of internal pressure close to the hydrostatic value for the mean specific gravity 0.3 of lung tissue . West (1966) indicates how this gradient is responsible also for certain variations of ventilation with height which as already remarked go some way towards match ing the bigger variations of perfusion now to be discussed.

With all reservations regarding p_A, we have seen that p_a and p_v show a much bigger vertical variation. In general, then, the lung can be divided into three zones, defined (West 1966) as follows:

Zone 1 is the highest zone where $p_a < p_A$.

Zone 2 is the intermediate zone where $p_v < p_A < p_a$.

Zone 3 is the lowest zone where $p_A < p_v$.

In Zone 1, any section of capillary (wherever its pressure may be between p_a and p_v) collapses under the alveolar pressure p_A and is without flow, as confirmed experimentally by West, Glazier, Hughes & Maloney (1969). In Zone 3 the tubes run fully open; low down in zone 3 they are observed to bulge slightly, which may by mechanisms discussed in Section 2.5 be responsible for the very low pressure drops $p_a - p_v$ there found.

It might be thought sufficient to say that Zone 2 is a transitional zone, but its behaviour is of some theoretical interest and has attracted much attention. For thin collapsible tubes carrying homogeneous fluid between reservoirs at pressures p_a and p_v an intermediate pressure p_A external to the tube produces (Permutt, Bromberger-Barnea & Bane, 1962) a partial closure at the distal end with a resulting local high resistance balancing an abrupt drop in fluid pressure from p_A to p_v. The flow times the normal resistance of the main tube equals the pressure drop $p_a - p_A$ up this distal constriction. This implies the flow property often described as a "Starling resistor" (a clearer designation than the above authors' name "waterfall" with its inertial connotations): flow is proportional to $p_a - p_v$ till p_v drops below p_A and thereafter is proportional to $p_a - p_A$. The distal partial collapse is inevitable for homogeneous fluid since full flow in the tube must cause tube closure wherever pressure falls to below p_A, but complete closure would allow the proximal pressure p_a to penetrate the whole tube and force it open.

The situation is more complicated for whole blood, since flow can cease in passages not fully closed and support a pressure gradient through dry friction of red blood cells against walls (Section 2.5). Observations indicate that the Starling resistor remains a good description of pressure-flow relations but that the mechanism underlying it changes (West, Glazier, Hughes & Maloney 1969): briefly, it involves closure of an increasing number of passages as p_v falls below p_A, and recruitment of additional passages as p_v rises again.

Note that during exercise p_a is increased so that Zone 1 (without flow) may be absent; indeed, all the zone boundaries are then raised. Note also that in this brief account of the pulmonary circulation we have not implied that the capillary passages have the same tubular geometry as in the systemic circulation. Recent opinion (see for example Fung & Sobin 1968) tends to a view of the different capillary passages in the septum as being like the alternative routes for threading one's way across a hall crowded densely with thick pillars! However this may be, most pulmonary capillaries are passages which tend to squeeze red cells smaller than their undistorted diameter of 8 microns. This facilitates gas transfer between the pulmonary membrane and the hemoglobin, but it does raise special questions of lubrication and resistance in very narrow capillaries, which are taken up in Section 2.5 after the problems of resistance in vessels not quite so narrow are first discussed in Section 2.4.

2.4. Axial Concentration

To complement discussions in Lecture 1 about several kinds of departure from Poiseuille's law within large vessels, the rest of Lecture 2 is devoted to those quite different kinds of departure that are found in very small vessels, and to their implications for the microcirculation. These are departures observed in passages whose width is comparable with the undistorted-red-cell diameter of 8 microns, and thay take a rather surprising form: as the diameter of tube falls below about 30 microns, the pressure drop starts to take values substantially less than those indicated by Poiseuille's law using viscosity figures found adequate for blood flow in large tubes. This phenomenon, discovered by Fahraeus & Lindquist (1931), is discussed in Section 2.4. However the reversed tendency (to be discussed in Section 2.5) is found in the narrowest capillaries, with widths around 6 microns or less. Accordingly, an "apparent viscosity" inferred from Poiseuille's law falls initially (by about 50%) as the tube diameter decreases, reaching a minimum at a diameter of around 12 microns, and then rises to well beyond the large-tube value as the diameter falls further.

These phenomena are not significantly dependent on any variation of whole-blood viscosity as a function of rate of shear, such as was mentioned under heading (iv) in Lecture 1; actually, in the microcirculation as in large vessels, we are concerned with rates of shear too large for this variation to be substantial. It is essentially the interaction of suspended red cells with the vessel wall that produces the Fahraeus-Lindquist effect, to be discussed in Section 2.4; considerations relevant to the reversal of the effect, occurring in passages so narrow that red cells need to suffer substantial deformation even to enter them, are postponed to Section 2.5.

Fahraeus & Lindquist (1931) found, in fact, a statistical tendency for red cells to shun the wall and to be concentrated preferentially toward the axis. We momentarily postpone discussing the subtle fluid mechanics involved in this observed "axial concentration" phenomenon, in order to give first the rather simple explanation of how it brings about the reduced apparent viscosity.

Axial Concentration

The kinematic viscosity of the blood plasma at body temperature is about $1.5 \text{ mm}^2 \text{ s}^{-1}$, while that of whole blood increases with increasing haematocrit (that is, concentration of red cells by volume) slowly at first but at an increasing rate; that is, the curve of kinematic viscosity against haematocrit is concave upwards, passing through values around 4 at normal haematocrits of 0.4 to 0.5, and then rising to much higher values. Under conditions of axial concentration, therefore, the viscosity varies greatly with radial position, from a value near 1.5 in the almost cell-free region within a few microns of the wall to far larger values near the axis.

On the other hand, the elementary theory of laminar axisymmetrical flow with viscosity μ dependent on distance r from the axis tells us that at given pressure gradient the equilibrium of a cylinder of radius r requires the rate of shear to vary as $\mu^{-1} r$; in the situation just described both factors combine to make the shear very much greater near the wall than elsewhere, as is observed. The total flow can be written as an integral with respect to r of the product πr^2 times the rate of shear, which is proportional to $\mu^{-1} r^3$. This r^3 factor weights still more the high values of μ^{-1} found where r is greatest; that is, near the wall. In fact we conclude from the analysis that the reciprocal apparent viscosity μ_{ap}^{-1} (proportional to flow divided by pressure gradient) is the weighted mean of μ^{-1} with respect to r, using r^3

as the weighting factor.

This very high weighting given to the high values of μ^{-1} present in any almost cell-free layer near the wall makes it easy to see why an apparent viscosity as low as 2 to 2.5 $mm^2 s^{-1}$ is typically observed for tubes of 10 to 30 microns diameter. Whithmore (1968) points out that even the extreme hypothesis of effectively infinite viscosity for values of r up to 80% of the tube radius, and constant finite viscosity beyond, gives an apparent viscosity only 1.7 times that constant value. This model might represent blood of high haematocrit filling a 30 micron tube except for a cell-free layer 3 microns thick near the wall. Alternatively, for a 10 micron tube it might represent the effect of individual red cells passing down the tube in single file, shunning the wall (see below) to form an "axial train" of diameter 8 microns within which plasma shearing is effectively prevented.

Postponing to Section 2.5 the question of what happens at still lower tube diameters when plasma layer thickness are far more restricted in value, we next note other consequences of axial concentration besides the Fahraeus-Lindquist effect. First, it means that the "average speed of red cells" exceeds the "average speed of plasma" in these vessels in the 10 to 30 micron size bracket. Hence, on the average, a red cell spends a smaller proportion of its time within such vessels than does an average speck of plasma! It follows (Whitmore

Axial Concentration 45

1968) that the haematocrit within such vessels is substantially less than that for the body as a whole. Such haematocrit reduction may be important in an organ like the kidney which has a process to carry out (Lecture 1, Section 1.11) upon the blood plasma alone. Indeed, in the kidney circulation, a further reduction of haematocrit is thought to occur (Pappenheimer & Kinter 1956) by "skimming" off plasma from near the walls of arterioles into the fine capillary network, which is then bypassed by the remaining cell-rich fluid.

The rest of this Section is devoted to the fluid dynamics of "why red cells shun walls"; that is, to the mechanism of axial concentration, not at all an easy subject. Several errors must be avoided: explanations relying upon inertial effects (including "Magnus effect"), which are totally negligible as explained in Section 2.1; explanations valid only for dilute solutions; and explanations that ignore the red cell's shape and its deformability.

Mason & Goldsmith (1969) give a good review of this difficult field, though they are obliged to admit that no proper account avoiding all these errors has been given. In dilute suspensions at low Reynolds number, disk-shaped bodies like red cells spend most of their time almost edge-on (that is, presenting a small angle of incidence) to a sheared flow, but they "flip over" (rotate quicly so as to present the opposite edge to the flow) at regular intervals of the order of the reciprocal of the rate of shear. If they do this very near a wall, there is no actual collision but high pressures in a thin lubricating layer of plasma very near the wall (compare Section 2.5 below) may bring about some repulsion of the cell away from it.

If the disks are rigid, then no axial concentration occurs except for this avoidance of immediate proximity to the wall. Nevertheless, even in dilute suspensions there is an additional mechanism of axial concentration if the particles are deformable. During the "flip-over" process, the fluid stresses on the disk tend to stretch it in the direction of the distal edge when that is in the higher-velocity fluid but to compress it when that is in the lower-velocity fluid. Radial movements inwards in the former situation then exceed corresponding movements outwards in the latter, and so migration towards the axis can occur.

At normal haematocrits it is far from clear whether this mechanism can operate, since each cell is endlessly jostled by neighbouring cells. Most experimenters have observed some degree of axial concentration _additional_ to the existence of a relatively cell-free layer a few microns thick associated with wall-shunning, but its cause remains uncertain. It seems possible that if the statistical effect on a red cell of many close interactions with neighbouring cells were better understood we would see that in an equilibrium situation (statistically speaking) the neighbouring cells on the higher-shear side (where we may suppose that mean cell rotation is faster or alternatively that "flip-over" is more frequent) produce a relatively intenser effect that has to be balanced by a greater density of neighbouring cells on the lower-shear side. With this hypothetical remark we must leave the disputed question of the mechanism of axial concentration.

2.5. Lubrication Problems in Very Narrow Capillaries

We conclude Lecture 2 by discussing blood flow through passages so narrow that the shapes of red cells must be substantially deformed even to enter them. For this discussion we need to begin with a more detailed description than before of the undeformed shape of the red blood cell (erythrocyte) and of its deformability.

The erythrocyte is a cell of an unusually simplified kind; it is without a nucleus, consists essentially of a more or less fluid mass of haemoglobin surrounded by a thin flexible cellular membrane, and appears to respond to mechanical stimuli in a completely passive manner (in contrast to phagocytes, larger "white blood cells" with a nucleus and exhibiting a range of highly active responses to stimuli, including responses involving enormous changes of shape). The undeformed red cell, described earlier as a disk of diameter about 8 microns, is actually "dimpled" on both sides near the centre of the disk; in other words, it is concave outwards in the region. Discussion of how the convexo-concave shape of the undeformed red cell comes about would be out of place here, but we may note that such a shape, being far easier to deform without change of volume or surface area than a purely convex shape, facilitates greatly the cell constrictions needed in the very narrow capillaries of the body.

Nevertheless, external stresses are required to deform the cell from its equilibrium shape. From the work of Rand & Burton (1964) and of Fung (1966) we can give 0.5 mb as the order of magnitude of stress required (in the form, for example, of differences in the external pressures applied to different parts of the surface) to produce really substantial degrees of deformation, comparable with the cell's dimensions. Note that this stress is far smaller than the excess pressures required to distend capillaries: even in the lung, the work of West, Glazier, Hughes & Maloney (1969) indicates the corresponding figure to be two orders of magnitude greater, while in systemic capillaries the discrepancy is bigger still (Fung 1966).

When blood flow in a very narrow systemic capillary is made visible under the microscope, we observe what these figures would lead us to expect: the red cells moving in single file along the tube accomodate to its dimensions not by distending the tube but by themselves being deformed. Large de formations are observed, and of several different kinds: particularly a "parachute" shape in capillaries that are not so extremely narrow (this is an almost axisymmetrical configuration, curved convexly to the direction of motion), and an "edge-on" configuration bent or even folded along a diameter to pass through still narrower tubes as "crêpe suzette"! Similar results are obtained in glass tubes of similar diameters (Hochmuth, Marple & Sutera 1970). No doubt a variety of deformed erythrocyte shapes appear also in flow through the so-called "capillary marsh" in the lung, described at the end of Section 2.3, and we must suppose in all these cases that stresses of the order of 0.5 mb act to effect the larger deformations (and rather less for the smaller deformations).

    Lighthill (1968, 1969) pointed out the relevance to these flows of "hydrodynamic lubrication theory": a theory developed for engineering purposes of analysing how one surface may be enabled to slide over another while a substantial force is transmitted between them, through the presence of a very thin oil film. The high velocities in engineering applications are matched by the high viscosities of the oil so that we have a low-Reynolds-number situation (as in the microcirculation); indeed, it is the stress distribution due to viscous action in the oil that transmits the force, and normally the oil-film thickness adjusts passively to the value required for such transmission. Only when the "required value" is "too small" in some sense (e.g. to avoid asperities on the two surfaces from engaging with one another, or to avoid some kind of breakdown of the fluid properties of the oil) can a kind of "seize-up" or suspension of sliding occur, with "hydrodynamic lubrication" replaced by "dry friction".

Lighthill argued that the stress which the wall of a very narrow capillary passage needs to transmit to a red cell sliding past it so that the latter may be sufficiently deformed to pass through must, similarly, be transmitted by viscous action in a thin lubricating layer or plasma; that normally the plasma-film thickness must adjust passively to the value required for such transmission; and that suspension of sliding may occur when this required value is "too small" in some sense. His work was taken considerably further by Fitz-Gerald (1969a, 1969b).

Most of the stress required to deform the cell is a radial stress, that must be transmitted by means of a distribution of excess pressure along the lubricating layer (a distribution associated with its thickness distribution), although some supplementary deformation may be provided by axial viscous stress (Fitz-Gerald 1969a) to cells passing in the "parachute" configuration through tubes not so extremely narrow. It is of course purely viscous considerations that bring about a variation of pressure along the lubricating layer, with a gradient which simple dimensional arguments (supported by lubrication theory) suggest should be proportional to $\mu_p U/h^2$: that is, proportional to the plasma viscosity μ_p and to the velocity of sliding U, and inversely proportional to the square of the plasma-film thickness h. Actually, film thicknesses of a few tenths of a micron are thus predicted as necessary to cause large deformations of red cells, at typical capillary flow speeds of 0.5 mms^{-1}, but the most significant conclusion from the theoretical observations is that the film thickness h should very roughly as $U^{1/2}$ (so that $\mu_p U/h^2$ has a roughly constant effect).

Evidently a substantial viscous force resisting red-cell motion, and given by Newton's law of viscosity as proportional to $\mu_p U/h$, is to be expected wherever h is very small. This demands, if h is proportional to $U^{1/2}$, an excess pressure drop in a capillary also proportional to $U^{1/2}$. The nonlinearity of these predictions, that both plasma-film thickness and the excess pressure drop that it produces vary as $U^{1/2}$, makes them interesting: the latter pressure drop, for example, must exceed any normal linear component (proportional to U) when U is small enough. A variety of detailed mathematical models developed by Lighthill and Fitz-Gerald show that these conclusions continue to be predicted by hydrodynamic lubrication theory under a wide range of assumptions (taking into account, for example, departures from axisymmetrical motion and also loss of fluid through gaps in the capillary wall), although none of the models allow for anything like an adequate detailed mechanics of the red-cell deformation.

Although ordinary microscopic studies of red-cell motion in capillaries do not give sufficient resolution to determine the thickness of the lubricating plasma film, Hochmuth, Marple & Sutera (1970) made some very careful observations of blood flow in glass micro-capillaries from which plasma-film thicknesses were obtained in generally good accord with the above considerations. For example, their figure 5 indicates that in tubes of diameters (a) 4.5, (b) 6.7 and (c) 8 microns the plasma film thickness at low velocities takes values the shape of whose dependence on U is close to $U^{1/2}$ (multiplied, if U is measured in mm/s, by (a) 0.6, (b) 1.0 and (c) 1.3 microns respectively), but that the curve flattens off to a constant thickness above about $U = 1$ mm/s (rather as if further increase of film thickness were prevented by a large increase in resistance by the cell to the extra deformations it would demand).

These results encourage one to study observations on peripheral resistance *in vivo* and look for a component of total pressure drop proportional to the square root of flow rate; such a component might represent an extra low-velocity effect of the blood's passage through very narrow capillaries, and be additional to the main term directly proportional to flow (possibly together, at very high flow rates, with a term proportional to flow squared associated with entry-region effects). Actually the literature of peripheral resistance is full of such data, in which the pressure-flow curve comes parabolically into the origin with a vertical tangent (Green, Rapela & Conrad 1963). In the past the data were interpreted differently (for example, in terms of increased dilatation of the microcirculation at increased perfusing pressure), but it seems more and more probable that enhanced pressure drop at low flows due to thinning of lubricating plasma films in the narrowest capillaries is the true explanation.

Thus, as tube diameter falls, the Fahraeus-Lindiquist effect should be thought of as predicting only a local minimum in apparent viscosity (Section 2.4), followed by a rise to well above the normal whole-blood viscosity at small tube diameters. This same result was observed by Dintenfass (1968) for flow of blood of haematocrit 0.49 through a very narrow gap between parallel plates (see his figure 7).

Pulmonary capillary flow (see the end of Section 2.3) may be intermediate in character between flow in tubes and such flow in parallel plates. Certainly, many of its characteristics can be readily interpreted by lubrication theory, including stoppage of flow in particular pathways when hydrodynamic lubrication breaks down. Recently, Dr J.B. West carried out computer simulations of capillary networks (unpublished) which indicated how the uneven filling of different pathways in the pulmonary circulation observed (Section 2.3) by him and his colleagues would be expected to be very marked on the assumption of the nonlinear resistance law here postulated.

Some confirmation that the pressure-flow relations in the lung are directly related to the deformability of red cells is provided by unpublished experiments of Dr Reginald Greene. Using red cells that had been made much less readily deformable by a heat treatment, he showed that the square-root-of-flow component in the overall pressure drop, present already for normal cells, was considerably increased for the treated cells, consistently with the idea from lubrication theory that the less readily deformable cells would need a much higher value of $\mu_p U/h^2$.

In the systemic circulation, the non-linear resistance law for very narrow capillaries may increase the effectiveness of vasomotor control of peripheral prefusion (Section 2.2), essentially since an increase of arteriolar resistance reduces the flow rate and thus automatically increases the capillary component of resistance. Note also that, in systemic capillaries, some "bunching" of red cells is observed. This may be interpreted (see Whitmore 1968) in terms of slight variations in red-cell size or constrictability, leading to slight variations in plasma-film thickness and thus in rate of leakback of plasma through the film. Plasma then tends to accumulate behind any unusually small or easily constrictable cell.

Detailed models of the plasma-flow streamlines between successive red cells in a very narrow capillary have been obtained, and applied to estimation of the combined effects of convection and diffusion of particular solutes (see Section 2.1) by Aroesty & Gross (1970) and by Fitz-Gerald (1971). They conclude that diffusion is dominant for dissolved gases like O_2 and CO_2 but that convection by a toroidal circulation in what Prothero & Burton (1961) called the "bolus" of viscous fluid between two cells, as well as in the flow leaking back through the lubricating plasma film, may be quite significant for transfer of low-diffusivity macromolecules in the systemic circulation.

All this wotk on lubrication problems in the microcirculation is at very early stage, but it seems to be a body of ideas likely to find increasing application to pro blems of real physiological interest as time goes on.

References

[1] Aroesty, J. & Gross, J. 1970 "Convection and Diffusion in the Microcirculation". Report RM-6214-NIH. Santa Monica, Calif.: The Rand Corporation.

[2] Barua, S. N. 1963 Quart. J. Mech. Appl. Math. 16, 61.

[3] Caro, C. G. 1966 J. Physiol. 185, 501.

[4] Caro, C. G., Fitz-Gerald, J. M. & Schroter, R. C. 1971 Proc. Roy. Soc. B, 177, 109.

[5] Caro, C. G., Foley, T. H. & Sudlow, M. F. 1970 J. Physiol. 207, 257.

[6] Dintenfass, L. 1968 "The Viscosity of Blood...", Article (pp. 197-210) in "Hemorheology" (ed. Copley, A. L.). London: Pergamon.

[7] Fahraeus, R. & Lindquist, T. 1931 Am. J. Physiol. 96, 562.

[8] Fargie, D. & Martin, B. W. 1971 Proc. Roy. Soc. A, 321, 461.

[9] Farhi, L. E. 1966 "Ventilation-perfusion Relationship and its role in Alveolar Gas Exchange", Article (pp. 148-197) in "Advances in Respiratory Physiology" (ed. Caro, C. G.). London: Edward Arnold.

[10] Fitz-Gerald, J. M. 1969a Proc. Roy. Soc. B, 147, 193.

[11] Fitz-Gerald, J. M. 1969b J. Appl. Physiol. 27, 921.

[12] Fitz-Gerald, J. M. 1971 "Plasma motions in narrow capillary flow", 51, 463. J. Fluid Mech.

[13] Fowler, K. T., West, J. B. & Pain, M. C. F. 1966 Resp. Physiol. 1, 88.

[14] Fry, D. L. 1968 Circulation Res. 22, 165.

References

[15] Fung, Y. C. 1966 Federation Proc. 25, 1761.

[16] Fung, Y. C. & Sobin, S. S. 1968 Federation Proc. 27, 578.

[17] Green, H. D., Rapela, C. E. & Conrad, M. C. 1963 "Resistance and Capacitance Phenomena in Terminal Vascular Beds", Article (pp. 935-960) in "Handbook of Physiology, Section 2: Circulation" (ed. Hamilton, W. F. & Dow, P.). Washington: American Physiological Society.

[18] Guyton, A. C. 1966 "Textbook of Medical Physiology", London: W. D. Sanders.

[19] Hochmuth, R. M., Marple, R. N. & Sutera, S. P. 1970 Microvascular Res. 2, 409.

[20] Jaffrin, M. Y. & Shapiro, A. H. 1971 Ann. Rev. Fluid Mech. 3, 13.

[21] Kraemer, K. 1967 Arch. f. Kreislaufforschung 52, 79.

[22] Lighthill, M. J. 1968 J. Fluid Mech. 34, 113.

[23] Lighthill, M. J. 1969 "Motion in Narrow Capillaries from the Standpoint of Lubrication Theory", Article (pp. 85-104) in "Circulatory and Respiratory Mass Transport" (ed. Wolstenholme, G.E.W. & Knight, J.). London: J. & A. Churchill.

[24] Lighthill, M. J. 1970 "Turbulence", Article (pp. 83-146) in "Osborne Reynolds and Engineering Science Today" (ed. McDowell, D.M. & Jackson, J.D.). Manchester University Press.

[25] Macklem, P.T. & Mead, J. 1966 J. Appl. Physiol. 22, 395.

[26] Mason, S.G. & Goldsmith, H.L. 1969 "The Flow Behaviour of Particulate Suspensions", Article (pp. 105-129) in "Circulatory and Respiratory Mass Transport" (ed. Wolstenholme, G.E.W. & Knight, J.). London J. & A. Churchill.

[27] McConologue, D.J. & Strivastava, R.S. 1969 Proc. Roy. A,

307, 37

[28] Pappenheimer, J.R. & Kinter, W.B. 1956 Am. J. Physiol. 185, 377.

[29] Pattle, R.E. 1966 "Surface Tension and the Lining of the Lung Alveoli", Article (pp. 83-105) in "Advances in Respiratory Physiology" (ed. Caro, C.G.). London: Edward Arnold.

[30] Pedley, T.J., Schroter, R.C. & Sudlow, M.F. 1970 Respiration Physiol. 9, 371.

[31] Prothero, J. & Burton, A.C. 1961 Biophys. J. 1, 565.

[32] Rand, R.P. & Burton, A.C. 1964 Biophys. J. 4, 115.

[33] Seed, W.A. & Wood, N.B. 1971 Cardiovascular Res. 5, 319.

[34] Taylor, M.G. 1966 Biophys. J. 6, 697.

[35] West, J.B. 1966 "Regional Differences in Blood Flow and Ventilation in the Lung.", Article (pp. 198-254) in "Advances in Respiratory Physiology" (ed. Caro, C.G.). London: Edward Arnold.

[36] West, J.B., Glazier, J.B., Hughes, J.M.B. & Maloney, J.E. 1969 "Pulmonary Capillary Flow, Diffusion, Ventilation and Gas Exchange", Article (pp. 256-276) in "Circulatory and Respiratory Mass Transport" (ed. Wolstenholme, G.E.W. & Knight, J.). London: J. & A. Churchill.

[37] Whitmore, R.L. 1968 "The Dynamics of Blood Flow in Capillaires", Article (pp. 77-87) in "Hemorheology" (ed. Copley, A.L.).

[38] Womersley, J.R. 1955, J. Physiol. 127, 553

Contents

		Page
Preface	....	3
Lecture 1.	General Survey	
1.1	Introduction............................	5
1.2	Steady Secondary Flow..................	11
1.3	Entry Regions..........................	12
1.4	Incipient Atheroma.....................	14
1.5	Distribution of Shear in Branched Systems.................................	15
1.6	Distribution of Resistance in Branched Systems.................................	18
1.7	Bronchial Resistance...................	20
1.8	Velocity Distributions in Pulsatile Flow...................................	21
1.9	Pulse Propagation......................	23
1.10	Turbulence in the Blood Stream.........	25
1.11	Urinary Tract..........................	27
Lecture 2.	The Microcirculation	
2.1	Introduction...........................	29
2.2	Vasomotor Control of Peripheral Perfusion....................................	33
2.3	Pulmonary Perfusion and Ventilation...	36
2.4	Axial Concentration....................	41
2.5	Lubrication Problems in Very Narrow Capillaries............................	47
References	....	56
Contents	....	59

If you have any concerns about our products,
you can contact us on
ProductSafety@springernature.com

In case Publisher is established outside the EU,
the EU authorized representative is:
**Springer Nature Customer Service Center GmbH
Europaplatz 3, 69115 Heidelberg, Germany**

Printed by Libri Plureos GmbH
in Hamburg, Germany